"Dr. Eliot opens our eyes scientifically to what should be obvious common sense—overachievers don't surpass all others by following normal behavior or thinking! Read Overachievement and unleash your untapped potential."

— Dr. Mae Jemison
Former NASA Astronaut

....

"There are lots of books that teach you how to get in touch with your inner self, and not much else. But read John Eliot's Overachievement and you'll have a playbook of principles that drive you to success."

— Reggie Williams
Former NFL Pro-Bowler and Disney Vice President

....

"At the highest levels of any profession, winning results often come down to who has the best mental game. John Eliot will show you how to prepare your mind so you can consistently come through in key situations."

— Howie Long
NFL Hall of Famer, Actor, and Fox Sports Broadcaster

OVERACHIEVEMENT

The REAL Story Behind What it Takes to be Exceptional

from SmarterComics™

by John Eliot, Ph.D.

Illustrated by Nathan Lueth

Adapted by Cullen Bunn

Executive Editor	**Creative Director**
Corey Michael Blake	Nathan Brown

rtc

A Round Table Companies Production
www.roundtablepress.com

FOREWORD
By Dr. John A. Jane, Sr.

Dr. Jane has been professor and chairman of the neurosurgery department at the University of Virginia since 1969, and has won the Cushing Medal, the highest honor granted by the American Association of Neurological Surgeons. An international figure in the field, Dr. Jane was the neurosurgeon who treated Christopher Reeve after the tragic 1995 horseback riding accident that left the actor a quadriplegic.

A decade ago, I was invited to sit on a panel at a conference on medical education. There was a large audience, perhaps 1,000 people in the room. One of the big controversies of the day in medicine was on the table – the long hours and stress that young interns and residents had to endure. The moderator asked each of us how we addressed the issue of stress in our training programs. My answer was: "I try to raise the stress on my residents to the highest possible level, and if they can't take it, I fire them."

The audience was stunned. Some people began clapping, but many in the room also booed. I was not surprised. Stress and single-minded commitment to work have been demonized not just in medicine but throughout the culture. And the anti-stress forces seem to have won, at least in medical education. Residents are now prohibited by law from working more than 80 hours, though a hospital can petition to extend the work week another eight hours. I wish John Eliot's book had been available during that conference and the subsequent wrangles over "over-work" among young doctors. This book is a welcome antidote to the conventional wisdom among mainstream psychologists that to perform well we must be "relaxed." The good surgeon will be keyed up and full of the kind of nervous energy that is absolutely necessary to spend the hours required to focus on a complicated surgery – and deal with every possible contingency.

John Eliot knows this. As a psychology academic and college teacher, he is familiar with the extraordinary progress that's been made in the field of neuroscience. Much of his own field research on "high performers" was done watching surgeons work and talking to them about how they thought under pressure (including me, I might add, in the interest of full disclosure.) He also spends considerable time these days advising athletes, musicians, and businesspeople on how to transform their talent and training into great work. Dr. Eliot understands why stress is a good thing. He also understands that to be good at surgery – and just about every other complicated career and task under the sun – requires not just training in various techniques but also training in how to use the mind. I agree. My job as a professor of neurosurgery is to teach accomplished and supremely talented young doctors how to think under the gun and in the face of extreme adversity. I do so by creating as tough an environment as possible. If you're training top gun pilots, you don't make things easy for them.

The best in my business also have high levels of what Dr. Eliot calls "over-confidence." I remember starting out, wanting to be a great neurosurgeon. I expect that same dream to be the number one priority in my students' lives, too.

Such single-minded commitment doesn't often sit well with friends, family, even colleagues. I can now recommend that the nay-sayers read the chapter in this book on how top performers across the board are inclined to "put all of their eggs in one basket." It's a prime ingredient in becoming the best at what you do. And, as Dr. Eliot explains using anecdotes from business, sports, entertainment, as well as medicine, when you are committed to what you do, when you love it, those 88 hour work weeks fly right by. One of the hardest parts of my job is to inform my residents who've worked through the night and are bumping up against their weekly time limit, that they have to go home. "And miss tomorrow's cases?" they ask. They may be tired, but they do not want to pass up any opportunity to increase their odds of becoming a great neurosurgeon.

To be sure, it's not the way "normal" people go at things. But – and this is central to Dr. Eliot's view of high performance – how do you become extraordinary at what you do by settling for what's normal? If normalcy is your aim in life, then Overachievement is not the book for you. I have been fortunate to work with some very talented people, and there has been nothing normal about them. Some have been my patients. When the actor Christopher Reeve was thrown from his horse at his Virginia farm, he was rushed to my hospital, where I led a team through the critical stabilization procedure. At the time, some did not think that we had done Reeve a favor, including Reeve himself during his darkest moments. But then, against all the scientific evidence to the contrary, he decided that he would devote himself to regaining the ability to walk. That became his all-consuming dream – and Christopher Reeve's dream has revolutionized spinal cord research. His mature realization of his own situation and his empathy for the suffering of everyone else with spinal cord injuries has provoked and inspired a national and international initiative to find a cure.

Reeve's overachievements remind me of a quotation from George Bernard Shaw that my partner, Dr. Neal Kassell, has on his office wall. It's so on the mark for John Eliot's view of performance that I'm surprised he missed it. Let it be my way of making a good book even better:

The reasonable man adapts himself to the world. The unreasonable one persists in trying to adapt the world to himself. All progress depends on the unreasonable man.

If you aspire to push things forward or be the best at your business, if you're wondering how you can maximize your talent, if you're eager for some insight into how the human mind works and why we have now entered what science has dubbed "The Millennium of the Brain," and if you have no problem in being perceived as a bit "unreasonable" or "abnormal," then keep reading.

INTRODUCTION

Greetings!

Thank you for picking up a copy of the Overachievement comic book.

Every once and again, when I am giving a keynote address for a company, I joke about writing. I tease that the toughest part about producing the original hardcover version of Over-achievement was posing for so many hours, modeling for the cover. In fact, the Superman-in-a-business-suit image has proven to be quite an effective teaching metaphor when I'm talking about self-confidence, self-perception, and bold seachange dreams.

But my days pretending to be disguised as Clark Kent are over. To make a wardrobe switch, I no longer have to ferret out phone booths (which, by the way, are really hard to find these days; talk about needing x-ray vision). It's no longer just allegory speak when I talk about being Superman. Thanks to the brilliant vision of Corey Blake and the awesome artwork of Nathan Lueth, I've officially joined the superhero ranks. The transformation and working with Round Table Companies and SmarterComics has been a blast.

Now, my hope is that you'll join me!

Over the course of the pages to follow, I will walk you through a world of overachievers. I'll share with you the ingredients of real genius—not 180 IQ, fast as a speeding bullet, genetic endowment, but rather the type of thinking available to all us mere mortals that can, truly, change our futures. And while the backdrop of this book may be futuristic, the skills and habits I will discuss are performance strategies you can use, today, to be better at what you do and better at the activities you cherish the most.

Be warned: I'm going to lay it out there. I'm going to be audacious, controversial even. I do so for the purpose of pushing you, and pushing your mind. After all, my work isn't about being comfortable. It's about striving for greatness. And you don't become great by resting on your laurels, following the safest path, or eliminating challenges. By definition, you have to get past the norm. You have to have a touch of eccentricity about you.

So if you're someone who wants to break barriers, who wants to find out what you're really made of, who wants to live a life of excellence… if you're someone who embraces Einstein's crazy hairdo, then please read on. And enjoy!

I look forward to hearing about your tales…

All the best,
Doc Eliot

OVERACHIEVEMENT

The REAL Story Behind What it Takes to be Exceptional

from SmarterComics™

2

ARMED WITH THEIR QUICK-FIX POTIONS, PSYCHOLOGISTS AND PERFORMANCE COACHES HAVE BEEN PROPAGATING WHAT I CALL THE MYTHS OF HIGH PERFORMANCE.

THIS KIND OF SELF-IMPROVEMENT BALDERDASH WILL DO NOTHING BUT RELEGATE YOU TO A CAREER IN MEDIOCRITY.

*USE YOUR HEAD

*RELAX

*KNOW YOUR LIMITS

*WRITE DOWN GOALS

*GIVE

*DON'T PUT ALL YOUR EGGS IN ONE BASKET

*DON'T BE COCKY

*BE A TEAM PLAYER

*MINIMIZE YOUR RISKS

OVERACHIEVERS DON'T THINK REASONABLY, SENSIBLY, OR RATIONALLY.

THIS BOOK IS AIMED AT PEOPLE WHO WANT TO MAXIMIZE THEIR POTENTIAL.

TO ACHIEVE YOUR MAXIMUM POTENTIAL, YOU MUST STRETCH YOUR IMAGINATION, CHALLENGE YOUR BELIEFS, IGNORE THE PLEAS OF PARENTS, COACHES, SPOUSES, AND BOSSES TO BE REALISTIC.

REALISTIC PEOPLE DO NOT ACCOMPLISH EXTRAORDINARY THINGS BECAUSE THEY ALLOW THE ODDS AGAINST SUCCESS TO STYMIE THEM. THE BEST PERFORMERS KNOW THERE IS A WAY TO BEAT THE ODDS.

INSTEAD OF LIMITING THEMSELVES TO WHAT'S PROBABLE, THE BEST WILL PURSUE THE HEART-POUNDING, EXCITING, REALLY BIG DIFFERENCE-MAKING DREAMS, SO LONG AS CATCHING THEM MIGHT BE POSSIBLE.

SO LISTEN UP.

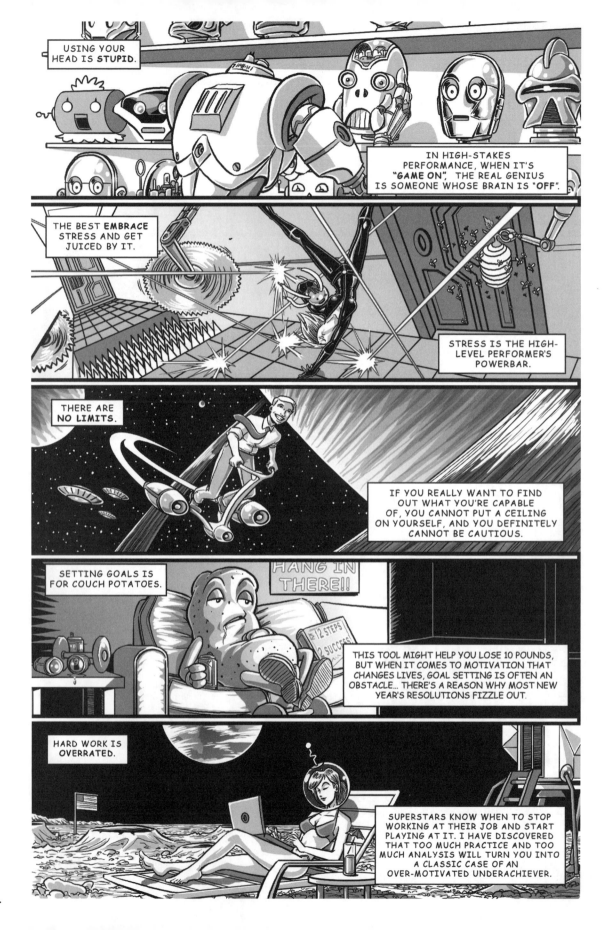

ALL THOSE EGGS BELONG IN ONE BASKET.

UNLIKELY ACCOMPLISHMENTS ARE BORNE OUT OF SINGLE-MINDED PURPOSEFULNESS.

METAPHORICALLY SPEAKING, ARROGANT S.O.B.S RUN THE WORLD.

A PERFORMER CAN NEVER HAVE TOO MUCH SELF-ASSURANCE. THE BEST IN EVERY FIELD ARE LIKELEY TO STRIKE MOST PEOPLE AS IRRATIONALLY CONFIDENT, BUT THAT'S HOW THEY GET TO THE TOP. WHEN PERFORMING, THEY ARE WILLING TO RISK OTHERS MISPERCEIVING THEM AS COCKY.

BEING THE OLD FASHIONED KIND OF TEAM PLAYER MAY GET YOU A GOLD STAR ON YOUR ANNUAL REVIEW OR AN "ATTA BOY" BUT IT MAY KEEP YOU FROM GETTING INTO THE CORNER OFFICE.

BY DEFENITION, STRIVING TO BE EXCEPTIONAL IS THE OPPOSITE OF FITTING IN WITH THE PACK.

LEGENDS RARELY SAY THEY'RE SORRY.

HAVING A LONG OR FREQUENT MEMORY FOR MISTAKES AND A SHORT OR INFREQUENT MEMORY FOR SUCCESS IS A GUARANTEED WAY TO DEVELOP FEAR OF FAILURE.

RISK-REWARD ANALYSIS IS FOR WIMPS.

GREATNESS DOESN'T HAPPEN WHEN YOU SPEND YOUR TIME TRYING TO ELIMINATE RISK. GREATNESS IS HOW YOU TACKLE DIFFICULTY, UNCERTAINTY, CRITICISM, AND CHALLENGE.

THE TRUSTING MINDSET

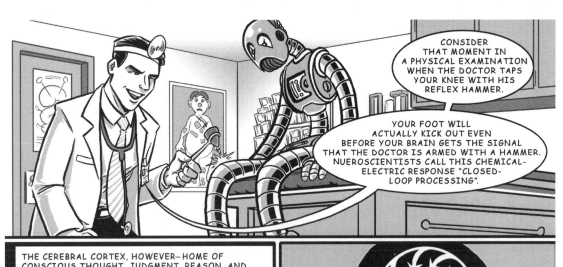

CONSIDER THAT MOMENT IN A PHYSICAL EXAMINATION WHEN THE DOCTOR TAPS YOUR KNEE WITH HIS REFLEX HAMMER.

YOUR FOOT WILL ACTUALLY KICK OUT EVEN BEFORE YOUR BRAIN GETS THE SIGNAL THAT THE DOCTOR IS ARMED WITH A HAMMER. NUEROSCIENTISTS CALL THIS CHEMICAL-ELECTRIC RESPONSE "CLOSED-LOOP PROCESSING".

THE CEREBRAL CORTEX, HOWEVER—HOME OF CONSCIOUS THOUGHT, JUDGMENT, REASON, AND CALCULATION—NEEDS BILLIONS OF NERVES TO DO ITS THING. INFORMATION PROCESSING THAT OCCURS ON THAT LEVEL IS CALLED OPEN-LOOP PROCESSING: OPEN, LITERALLY, TO INTERPRETATION.

WE HUMANS CAN ASSURE A KIND OF CLOSED-LOOP PROCESSING BY TAKING OUR CEREBRAL CORTEX OUT OF THE GAME, AND ALLOWING OURSELVES TO REACT TO SENSORY STIMULI WITH MOTOR RESPONSES WE HAVE ALREADY STORED.

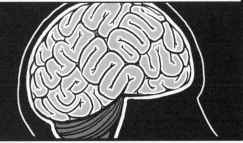

A STAR BASKETBALL PLAYER LOOKS AT THE RIM AND SHOOTS.

NO EVALUATING THE DISTANCE, NO DECISIONS ABOUT HOW HIGH TO EXTEND THE SHOOTING ARM OVER A DEFENDER OR HOW MUCH TO FLICK THE WRIST FOR PERFECT ROTATION. NO THINKING PERIOD.

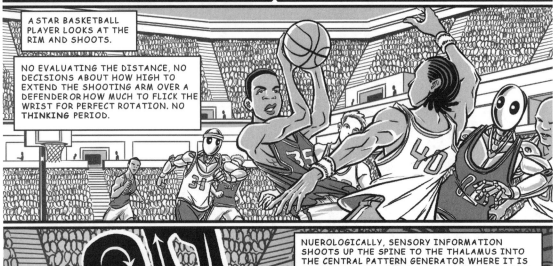

NUEROLOGICALLY, SENSORY INFORMATION SHOOTS UP THE SPINE TO THE THALAMUS INTO THE CENTRAL PATTERN GENERATOR WHERE IT IS ORGANIZED AND SENT BACK TO CAUSE THE ARMS AND HANDS TO DO WHAT A BASKETBALL PLAYER HAS TRAINED THE ARMS AND HANDS, THOUSANDS OF TIMES, TO DO.

FOR THE STAR BASKETBALL PLAYER, IT'S AS INSTINCTIVE AS IT IS FOR A SQUIRREL, EXECUTED THE SAME WAY AS TOSSING A SET OF KEYS…

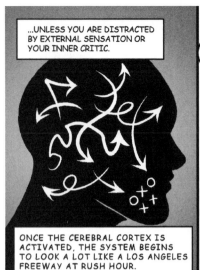

...UNLESS YOU ARE DISTRACTED BY EXTERNAL SENSATION OR YOUR INNER CRITIC.

ONCE THE CEREBRAL CORTEX IS ACTIVATED, THE SYSTEM BEGINS TO LOOK A LOT LIKE A LOS ANGELES FREEWAY AT RUSH HOUR.

WHEN THE CEREBRAL CORTEX IS INVOLVED, THE BRAIN'S PATTERN GENERATORS GET OVERLOADED AND THE SYSTEM GETS BOGGED DOWN, PRODUCING LESS EFFICIENT, LESS SUCCESSFUL ACTION, WITH A GREATER NUMBER OF MISTAKES.

TOSSING A SET OF KEYS SEEMS TO REQUIRE NO THOUGHT; IT'S VERY SQUIRREL-LIKE. THE CONSEQUENCES ARE MINIMAL, SO WE DON'T BOTHER TO USE OUR CEREBRAL CORTEX.

BUT IF I TOLD A LARGE GROUP TO COME BACK NEXT WEEK FOR ONE CHANCE TO TOSS THAT SAME SET OF KEYS INTO MY HAND—THIS TIME FOR A $1 MILLION PRIZE—ENTER OPEN LOOP PROCESSING.

THINGS WOULD TURN SCIENTIFIC AND PEOPLE WOULD START OVER-PRACTICING.

ONCE THE PRESSURE IS ON, PEOPLE TRY TO TOSS A SET OF KEYS ACROSS THE ROOM AND END UP CHOKING.

SUPERSTARS DON'T THINK THAT WAY. WHEN IT'S GO TIME, WHEN IT REALLY COUNTS, TECHNIQUE IS NOT ON THEIR MINDS.

WHEN A JOB HANGS IN THE BALANCE, GREAT THINKERS RESIST THE URGE TO BE SMART, CAUTIOUS, OR SCIENTIFIC.

FOR THEM, PERFORMANCE IS SIMPLY "CHILD'S PLAY" WHICH SUGGESTS A USEFUL DEFINITION OF THE SUPERSTAR'S EDGE:

THE TRUSTING MINDSET IS WHAT YOU USED **BEFORE** YOU KNEW ANY BETTER.

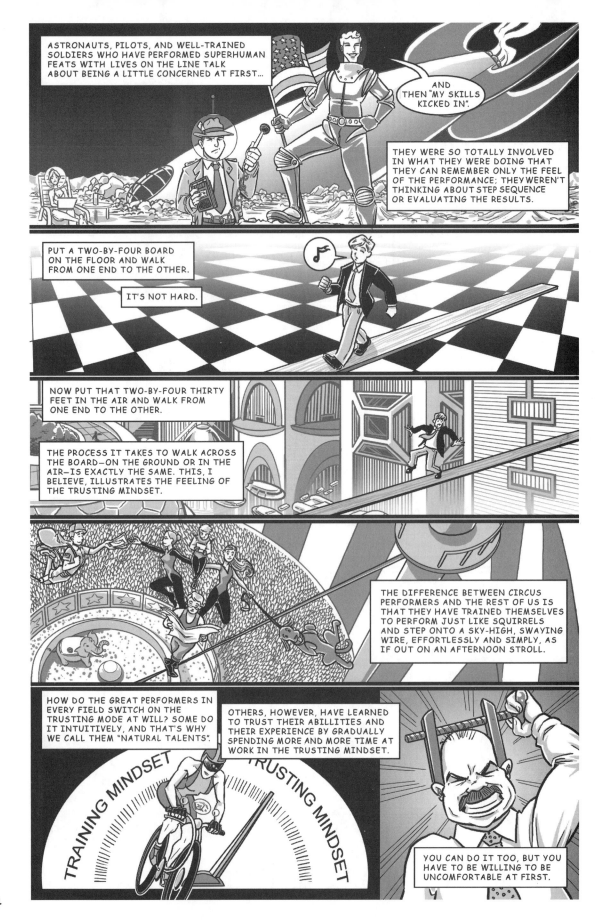

THRIVING UNDER PRESSURE

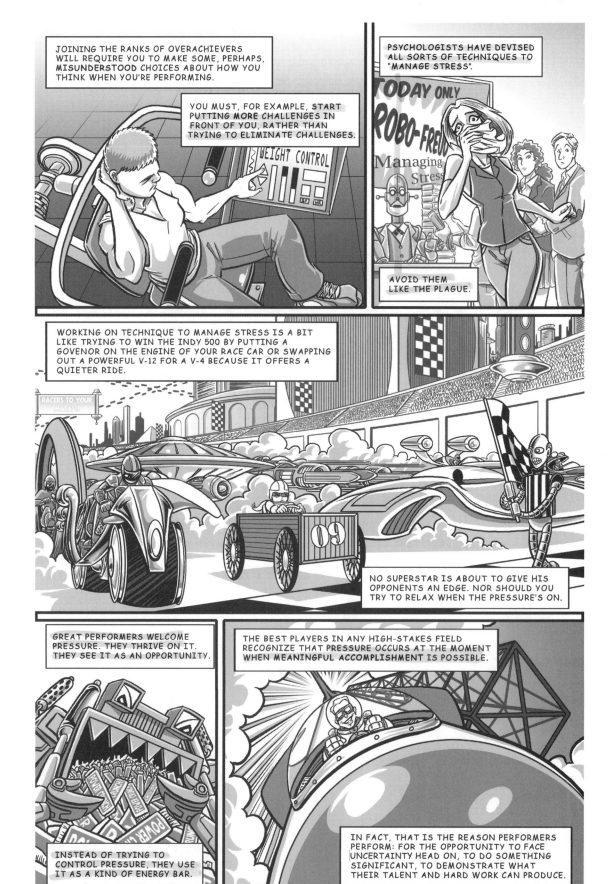

JOINING THE RANKS OF OVERACHIEVERS WILL REQUIRE YOU TO MAKE SOME, PERHAPS, MISUNDERSTOOD CHOICES ABOUT HOW YOU THINK WHEN YOU'RE PERFORMING.

YOU MUST, FOR EXAMPLE, START PUTTING **MORE** CHALLENGES IN FRONT OF YOU, RATHER THAN TRYING TO ELIMINATE CHALLENGES.

WEIGHT CONTROL

PSYCHOLOGISTS HAVE DEVISED ALL SORTS OF TECHNIQUES TO "MANAGE STRESS".

TODAY ONLY
ROBO-FREUD
Managing Stress

AVOID THEM LIKE THE PLAGUE.

WORKING ON TECHNIQUE TO MANAGE STRESS IS A BIT LIKE TRYING TO WIN THE INDY 500 BY PUTTING A GOVENOR ON THE ENGINE OF YOUR RACE CAR OR SWAPPING OUT A POWERFUL V-12 FOR A V-4 BECAUSE IT OFFERS A QUIETER RIDE.

RACERS TO YOUR MARKS!

09

NO SUPERSTAR IS ABOUT TO GIVE HIS OPPONENTS AN EDGE. NOR SHOULD YOU TRY TO RELAX WHEN THE PRESSURE'S ON.

GREAT PERFORMERS WELCOME PRESSURE. THEY THRIVE ON IT. THEY SEE IT AS AN OPPORTUNITY.

THE BEST PLAYERS IN ANY HIGH-STAKES FIELD RECOGNIZE THAT PRESSURE OCCURS AT THE MOMENT WHEN *MEANINGFUL ACCOMPLISHMENT* IS POSSIBLE.

INSTEAD OF TRYING TO CONTROL PRESSURE, THEY USE IT AS A KIND OF ENERGY BAR.

IN FACT, THAT IS THE REASON PERFORMERS PERFORM: FOR THE OPPORTUNITY TO FACE UNCERTAINTY HEAD ON, TO DO SOMETHING SIGNIFICANT, TO DEMONSTRATE WHAT THEIR TALENT AND HARD WORK CAN PRODUCE.

THEY KNOW THAT THEIR ABILITY TO PERFORM *CONSISTENTLY* WELL HAS NOTHING TO DO WITH IMAGINING A PEACEFUL ISLAND, REMINDING THEMSELVES TO BE CALM OR EMOTIONLESS, OR ELIMINATING JITTERS.

THEY DON'T WANT TO RELAX. FOR THEM, PRESSURE IS THE DOORWAY TO SUCCESS.

SHOW THEM SOMEONE LYING ON THE FLOOR WITH HIS EYES CLOSED, TRYING TO MAKE THE NERVES GO AWAY, AND THEY'LL SHOW YOU SOMEONE WHO IS EASY TO BEAT.

AT HIGH LEVELS OF BUSINESS, MEDICINE, ENTERTAINMENT, AND SPORTS, LEARNING HOW TO BRING DOWN YOUR PHYSIOLOGY WHEN THE PRESSURE IS ON WILL NOT IMPROVE PERFORMANCE. BEING A CLUTCH PLAYER MEANS *THRIVING* UNDER PRESSURE— WELCOMING IT, ENJOYING IT, MAKING IT AN ADVANTAGE YOU HAVE OVER YOUR COMPETITORS.

I CAN TEACH YOU HOW TO DO THIS, BUT FIRST YOU'LL HAVE TO RESTRAIN SOME ASSESSMENT, AND THAT WILL REQUIRE UNDERSTANDING TWO THINGS:

EVERYTHING YOUR BODY DOES IN INSTINCTIVE RESPONSE TO PRESSURE IS *GOOD* FOR PERFORMANCE.

PRESSURE IS DIFFERENT FROM ANXIETY; NERVOUSNES IS DIFFERENT FROM WORRY.

LIKE ALMOST EVERY ANIMAL, HUMANS HAVE BIMODAL *SYMPATHETIC* AND *PARASYMPATHETIC* NERVOUS SYSTEMS THAT HAVE EVOLVED OVER THOUSANDS OF YEARS.

SWEAT FLOWS—A SAFETY MECHANISM TO PREVENT THE BODY FROM OVERHEATING.

HANDS, FEET, OR KNEES BEGIN SHAKING. THAT'S THE BODY SENDING FASTER MOTOR SIGNALS FROM THE CORTEX THROUGH THE MOTOR NEURONS OUT TO THE EXTREMITIES SO REACTION TIME CAN BE QUICKER.

EYES DILATE, AND VISION BECOMES MORE ACUTE.

THE MIND FEELS LIKE IT IS RACING, MEANING IT IS PROCESSING A GREATER AMOUNT OF INFORMATION IN A SHORTER AMOUNT OF TIME.

AIRLOCK RELEASE

ALL OF THESE ADAPTATIONS ARE THE BODY'S WAY OF MAKING US PERFORM **MORE EFFICIENTLY** AND **MORE EFFECTIVELY** WHEN WE ARE UNDER THE GUN.

TRADITIONAL RELAXATION THINKING TEACHES YOUR MUSCLES TO LOSE TONE, YOUR BRAIN TO BE PASSIVE.

YOU CANNOT WIN GOLD MEDALS WITHOUT MUSCLE TONE, NOR CAN YOU PERFORM YOUR UTMOST WITH OTHER PARTS OF YOUR SYMPATHETIC NERVOUS SYSTEM SWITCHED TO "SLOW".

THE PHYSICAL SYMPTOMS OF FIGHT OR FLIGHT ARE WHAT THE BODY HAS LEARNED OVER THOUSANDS OF YEARS TO OPERATE AT A HIGHER LEVEL WHEN SITUATIONS ARE CRITICAL

CHOMP MUNCH SLURP RUNCH

ANXIETY, ON THE OTHER HAND, IS A COGNITIVE INTERPRETATION OF THE FIGHT OR FLIGHT RESPONSE.

MOST PEOPLE HAVE COME TO BELIEVE THAT ANXIETY AND STRESS GO HAND-IN-HAND. THAT ASSUMPTION, HOWEVER, IS DEAD WRONG.

STRESS NEED NOT PRODUCE ANXIETY.

BUTTERFLIES, COTTON MOUTH, AND A POUNDING HEART MAKE THE FINEST PERFORMER SMILE—THE SMILE OF A PERSON WITH AN ACE UP THEIR SLEEVE.

FIGHT OR FLIGHT SYMPTOMS COMPRISE THE EXTRA JUICE THEY'LL NEED TO GO UP AGAINST THE BEST, SO THEY WELCOME IT.

IN OUR CULTURE FEW WORDS CARRY MORE NEGATIVE CONNOTATION THAN "PRESSURE" AND "STRESS". STRESS GETS BLAMED FOR EVERY-THING THAT DOESN'T OTHERWISE HAVE A CLEAR DIAGNOSIS.

BUT STRESS IS NOT THE CAUSE; IT'S HOW YOU **INTERPRET** STRESS THAT CAUSES PSYCHOSOMATIC ILLNESS.

YOU NOW HAVE TO START TRAINING YOURSELF TO ACCEPT THAT AROUSAL IS A GOOD THING.

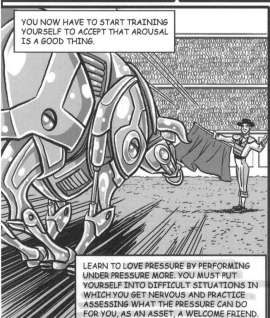

LEARN TO LOVE PRESSURE BY PERFORMING UNDER PRESSURE MORE. YOU MUST PUT YOURSELF INTO DIFFICULT SITUATIONS IN WHICH YOU GET NERVOUS AND PRACTICE ASSESSING WHAT THE PRESSURE CAN DO FOR YOU, AS AN ASSET, A WELCOME FRIEND.

OUR GREATEST HURDLE SIGNALS OUR GREATEST OPPURTUNITY TO EXCEL—TO TEST OUR MINDS, TO SEE HOW EXCEP-TIONAL OUR THINKING CAN BE.

RECOGNIZE THE ASSOCIATION BETWEEN THE NERVES AND THE POTENTIAL TO PERFORM, AS THE OLYMPIC CREED SAYS, "HIGHER, FASTER, STRONGER".

BREAKING BAD HABITS TAKES TIME. IN FACT, THE RESEARCH ON WHAT IT TAKES TO BREAK AN OLD HABIT AND LEARN A NEW ONE INDICATES THAT SUCH A TRANSITION COULD TAKE **THOUSANDS** OF TRIALS. SO THROW AWAY ALL THE QUICK FIXES AND GET TO WORK TRAINING YOUR MIND! THAT'S WHY MASTER MARTIAL ARTISTS FOCUS ON PUSHING THE LIMITS OF THEIR THINKING MORE THAN IMPROVING THEIR TECHNIQUE.

REMEMBER: THE ONLY TIME TOP PERFORMERS GET WORRIED IS WHEN THEIR HEART IS **NOT** RACING. UNLESS YOU LEARN TO LOVE PRESSURE—TO PERCIEVE STRESS AS AN ADVANTAGE—YOU ARE UNLIKELY TO ENTER THE RANKS OF THE EXCEPTIONAL. HALL OF FAME ATHLETE AFTER HALL OF FAME ATHLETE SAY THE SAME THING: "THE DAY I'M NO LONGER NERVOUS IS THE DAY I DECIDE TO RETIRE."

CREATING YOUR OWN REALITY

EXCEPTIONAL THINKERS SEE THE WORLD THROUGH THEIR OWN LENS.

IN FACT, THEY INVENT THE LENS. AND IF THAT LENS DOESN'T HELP THEM SEE THE WORLD THE WAY THEY WANT TO SEE IT, THEY INVENT ANOTHER.

WE HAVE TO UNDERSTAND THAT AS STRANGE AND OUTRAGEOUS AS SUCH PEOPLE OFTEN SEEM TO BE, IT IS THEY WHO MAKE US SIT BACK AND MARVEL AT HUMAN INGENUITY AND TALENT.

AS THE APPLE COMPUTER COMMERCIAL SAYS, PEOPLE LIKE JIM HENSON, PAUL NEWMAN, STEVE JOBS ARE "THE MISFITS, THE ODDBALLS, THE CRAZY ONES ... AND THEY ARE THE ONES WHO CHANGE THE WORLD."

MANY PEOPLE TEND TO CONSIDER THE WAY THEY THINK TO BE GENETICALLY DETERMINED, LIKE THE COLOR OF THEIR EYES OR HAIR.

IF YOU JUDGE YOUR IDENTITY ACCORDING TO HOW OTHER PEOPLE VIEW YOU, IF YOU DON'T BELIEVE YOU CAN MAKE A SHIFT IN YOUR THOUGHTS WHENEVER YOU WANT, IF YOU SHY AWAY FROM MISTAKES, THINKING THEY WILL DEFINE YOU—IF THAT'S THE WAY YOU TOLD YOURSELF YOUR BRAIN WORKS, THEN YOUR MINDSET HAS BECOME A MAJOR OBSTACLE TO BEING A SUCCESSFUL PERFORMER.

FORGET THE AGE OLD NATURE VERSUS NURTURE DEBATE.

EVERYONE CAN CHOOSE TO CHANGE HOW THEY THINK. BUT VERY FEW PEOPLE DO SO, BELIEVING THAT THE MIND IS "JUST WIRED THAT WAY" OR A PRODUCT OF HOW OUR PARENTS AND SCHOOL MOLDED US... OR THEY STAY STUCK BECAUSE OF SOME CLAIMED RESPONSIBILITY TO "THINK LIKE A GROWNUP".

WHAT YOU EXPERIENCE AS A FIXED OR HARD TO CHANGE MINDSET IS NOTHING MORE THAN A STRONG PATTERN OF SYNAPTIC JUNCTIONS WITH A LOT OF NEUROTRANSMITTERS. IF YOU WANT TO CHANGE THE WAY YOU THINK, YOU NEED TO WEAKEN THE OLD SYNAPTIC JUNCTION AND STRENGTHEN NEW ONES.

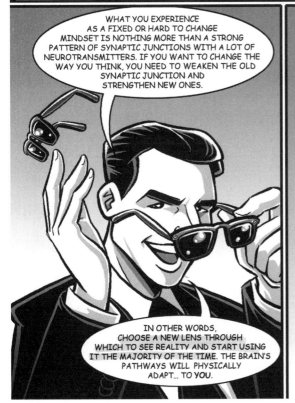

IN OTHER WORDS, CHOOSE A NEW LENS THROUGH WHICH TO SEE REALITY AND START USING IT THE MAJORITY OF THE TIME. THE BRAIN'S PATHWAYS WILL PHYSICALLY ADAPT... TO YOU.

MOST PEOPLE WALK AROUND WITH CONSTANT CONVERSATION GOING ON IN THEIR HEADS, LOGGING LITERALLY THOUSANDS OF THOUGHTS A DAY. MANY PEOPLE TREAT THIS INNER MONOLOGUE AS IF IT WERE A PHYSICAL NECESSITY OF BEING HUMAN. OR THEY DON'T EVEN PAY ATTENTION TO THE FACT IT'S GOING ON!

THIS IS THE CEREBRAL CORTEX DOING ITS THING. IT IS OUR SUPEREGO WARNING US TO RESPECT THE LAW AND THE VALUES OF THE COMMUNITY. THE HIGHER BRAIN AND A VALUE SYSTEM ARE VALUABLE THINGS TO HAVE WHEN YOU'RE STRATEGIZING IMPORTANT DECISIONS AND JUDGMENTS THAT AFFECT THE LIVES OF OTHERS.

DREAMS VS. GOALS

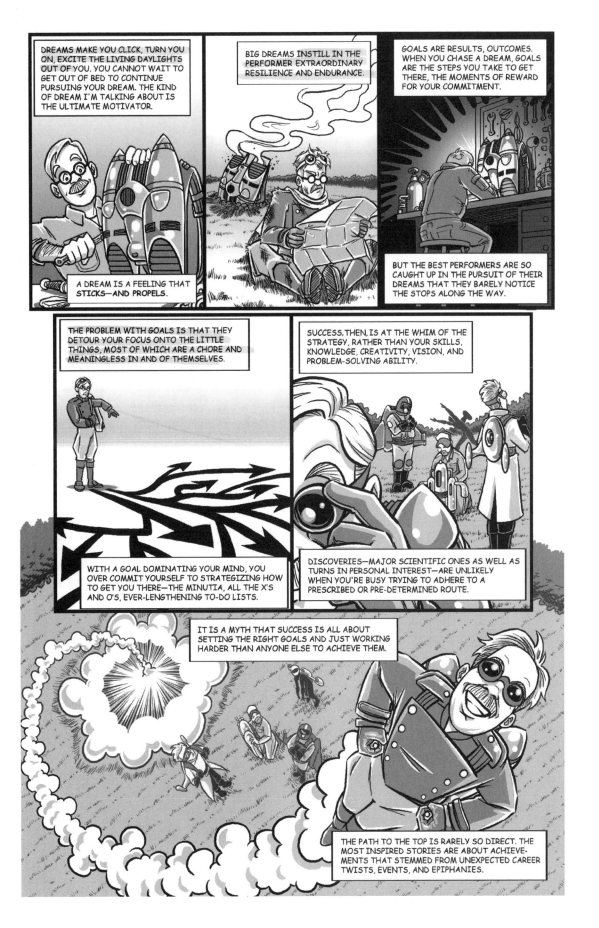

DREAMS MAKE YOU CLICK, TURN YOU ON, EXCITE THE LIVING DAYLIGHTS OUT OF YOU. YOU CANNOT WAIT TO GET OUT OF BED TO CONTINUE PURSUING YOUR DREAM. THE KIND OF DREAM I'M TALKING ABOUT IS THE ULTIMATE MOTIVATOR.

A DREAM IS A FEELING THAT STICKS—AND PROPELS.

BIG DREAMS INSTILL IN THE PERFORMER EXTRAORDINARY RESILIENCE AND ENDURANCE.

GOALS ARE RESULTS, OUTCOMES. WHEN YOU CHASE A DREAM, GOALS ARE THE STEPS YOU TAKE TO GET THERE, THE MOMENTS OF REWARD FOR YOUR COMMITMENT.

BUT THE BEST PERFORMERS ARE SO CAUGHT UP IN THE PURSUIT OF THEIR DREAMS THAT THEY BARELY NOTICE THE STOPS ALONG THE WAY.

THE PROBLEM WITH GOALS IS THAT THEY DETOUR YOUR FOCUS ONTO THE LITTLE THINGS, MOST OF WHICH ARE A CHORE AND MEANINGLESS IN AND OF THEMSELVES.

WITH A GOAL DOMINATING YOUR MIND, YOU OVER COMMIT YOURSELF TO STRATEGIZING HOW TO GET YOU THERE—THE MINUTIA, ALL THE X'S AND O'S, EVER-LENGTHENING TO-DO LISTS.

SUCCESS, THEN, IS AT THE WHIM OF THE STRATEGY, RATHER THAN YOUR SKILLS, KNOWLEDGE, CREATIVITY, VISION, AND PROBLEM-SOLVING ABILITY.

DISCOVERIES—MAJOR SCIENTIFIC ONES AS WELL AS TURNS IN PERSONAL INTEREST—ARE UNLIKELY WHEN YOU'RE BUSY TRYING TO ADHERE TO A PRESCRIBED OR PRE-DETERMINED ROUTE.

IT IS A MYTH THAT SUCCESS IS ALL ABOUT SETTING THE RIGHT GOALS AND JUST WORKING HARDER THAN ANYONE ELSE TO ACHIEVE THEM.

THE PATH TO THE TOP IS RARELY SO DIRECT. THE MOST INSPIRED STORIES ARE ABOUT ACHIEVE-MENTS THAT STEMMED FROM UNEXPECTED CAREER TWISTS, EVENTS, AND EPIPHANIES.

YOU KNOW THE TYPE: THE FIRST PERSON AT THE OFFICE, THE LAST TO LEAVE, NEVER TAKES A BREAK, GRINDS RIGHT THROUGH LUNCH, ALWAYS EAGER TO TAKE ON A NEW PROJECT, CONSTANTLY WORKING ON SKILLS AND SELF IMPROVEMENT, MONTHS OF UNUSED VACATION AND PERSONAL DAYS PILED UP.

THIS IS THE KIND OF EAGER BEAVER WHO IS SO COMMITTED THAT THEY SACRIFICE PERSONAL AND FAMILY TIME FOR THE ORGANIZATION, FOR A FANCIER TITLE, FOR A FEW EXTRA DOLLARS. THEY ARE IN IT FOR THE LONG TERM. THEY GIVE 110%.

OUR CULTURE LOVES THESE PEOPLE, CONSIDERING THEM THE VERY DEFINITION OF THE WORD DEDICATION.

AS OUR ESCALATING RATES OF HEART DISEASE AND PROZAC PRESCRIPTIONS SHOW, ALL THIS SACRIFICE ISN'T MAKING THE WORLD A BETTER PLACE.

THOSE WHO PLACE TOO MUCH VALUE ON THE NOTION OF LABORING HARDER THAN EVERYONE ELSE CAN BE DOING THEMSELVES—NOT TO MENTION THEIR REAL AMBITIONS, THEIR SELF-CONFIDENCE, AND THEIR FAMILIES—A DISSERVICE.

GOOD JOB, ROB

SUCH PERFORMERS HAVE BECOME THE EMBODIMENT OF THE CHERISHED AMERICAN RAGS TO RICHES BELIEF THAT SACRIFICE AND HARD WORK OPENS DOORS.

THE KIND OF PERSON WHO'S ALWAYS TENDING TO "JUST ONE MORE THING" BEFORE CALLING IT A DAY—THE PERSON THAT SOME MIGHT CALL A "GRUNT"—CAN BE A LIABILITY.

THERE IS A BIG DIFFERENCE BETWEEN HARD WORK AND GREAT WORK.

AND WHILE AMERICANS LOVE THE PERSON WHO GIVES IT HIS ALL AND ARE CRITICAL OF A NATURALLY GIFTED PERFORMER WHO MAKES IT LOOK EASY, AS COACH, CEO, OR HEAD OF SURGERY, I WILL TAKE THE SEEMINGLY **NATURAL PERFORMER** OVER THE GRINDER EVERY TIME.

COMMITMENT = CONFIDENCE

WHEN PEOPLE WARN, "DON'T PUT ALL YOUR EGGS IN ONE BASKET", EXCEPTIONAL THINKERS LAUGH THEM OFF; PUTTING ALL THEIR EGGS IN ONE BASKET IS A SECRET TO THEIR SUCCESS.

THEY DECIDE WHAT THEY WANT OUT OF LIFE AND WORK AND COMMIT TO THAT CHOICE WITH A SINGLE-MINDED FOCUS WE USUALLY IDENTIFY ONLY WITH STARVING ARTISTS WHO ENDURE POVERTY AND SCORN OF POPULAR TASTE TO PURSUE THEIR ARTISTIC PASSIONS.

THE BEST IN EVERY FIELD DO WHAT THEY DO BECAUSE ...

TO OUTSIDERS, THEIR BEHAVIOR CAN SEEM OBSESSIVE, MONOMANIACAL, OVERLY RISKY, OR DOWNRIGHT CRAZY. HEAVEN FORBID YOU DON'T FOLLOW THE OLD FASHIONED ADAGE OF HAVING A BACK-UP PLAN!

THEY SIMPLY CANNOT IMAGINE THEMSELVES DOING ANYTHING ELSE. IT'S LOVE THAT WINS THE DAY, NOT SOME SOCIETAL DRIVEN NOTION OF "SUCCESS".

BETWEEN 1661 AND 1666, ISAAC NEWTON SAT ALONE IN HIS STUDY AT CAMBRIDGE UNI-VERSITY DESCRIBING THE LAWS OF PHYSICS AND THE MATHEMATICS TO PROVE THEM.

HE THEN PROCEEDED TO SHOW HOW THE PHENOMENA OF THE REAL UNIVERSE, SUCH AS THE MOVEMENTS OF THE SUN AND THE MOON AND THE RISE AND FALL OF TIDES, ALL DANCE TO HIS NEW TUNE.

THE RESULTS WERE AN EXTRAORDINARY FEAT OF GENIUS, OVERTURNING THE ARISTOTELIAN COSMOLOGY THAT HAD DOMINATED HUMAN THOUGHT FOR ALMOST 2000 YEARS. NEWTON WAS 24 YEARS OLD!

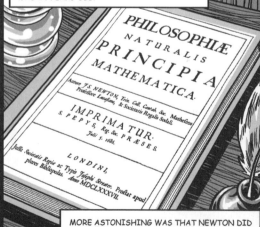

PHILOSOPHIÆ
NATURALIS
PRINCIPIA
MATHEMATICA.

Autore IS NEWTON, Trin. Coll. Cantab. Soc. Mathescos Professore Lucasiano, & Societatis Regalis Sodali.

IMPRIMATUR.
S. PEPYS, Reg. Soc. PRÆSES.
Julii 5. 1686.

LONDINI,
Jussu Societatis Regiæ ac Typis Josephi Streater. Prostat apud plures Bibliopolas. Anno MDCLXXXVII.

MORE ASTONISHING WAS THAT NEWTON DID NOT PUBLISH HIS FINDINGS FOR ANOTHER 20 YEARS—ONLY AFTER A FRIEND WORE HIM OUT, BEGGING HIM TO SHARE THIS KNOWLEDGE.

NEWTON TURNED HIS PERSONAL NOTES INTO HIS MASTERPIECE, THE PRINCIPIA MATHEMATICA, BUT ONLY DURING BREAKS FROM HIS NEW INTELLECTUAL OBSESSION, ALCHEMY, THE SPURIOUS SCIENCE OF TURNING IRON INTO GOLD.

COMMITMENT CAN PRODUCE **PRINCIPIA MATHEMATICA**. IT CAN ALSO PRODUCE BALDERDASH.

THE DESTINATION IS NOT THE POINT; THE **JOURNEY** IS WHAT THE GREATEST PERFORMERS LOVE. AND THEY GO AFTER IT WITH AN INTENSITY THAT MOST VIEW AS NUTS, OR EVEN IRRESPONSIBLE.

TO SUCCEED, PARTICULARLY EARLY IN LIFE, YOU HAVE TO GRAB ONTO A FEELING THAT WILL SEPARATE YOU FROM THE HERD, AN INTRINSIC, PASSIONATE LOVE FOR SOMETHING, AND GO FOR IT, ALL OUT.

YOU REALLY DO HAVE TO SPEND A LOT OF TIME PUTTING ALL YOUR EGGS IN ONE BASKET.

SOME MIGHT ALSO CALL IT "PUTTING BLINDERS ON". I CALL IT BEING SINGLE-MINDED IN PURPOSE, HAVING A TRUE MEANING TO YOUR LIFE—THE KIND OF MINDSET THAT TYPIFIES EXCEPTIONAL THINKERS.

MY CLIENTS OFTEN WONDER WHETHER THEY SHOULD HAVE A BACKUP PLAN IN CASE THEY FALL SHORT OF ACHIEVING THEIR DREAMS.

PLAN **A**

PLAN **B**

THINKING ABOUT CONTINGENCY PLANS EVEN BEFORE YOU BEGIN TO CHASE YOUR DREAM SHOWS A CERTAIN LACK OF CONFIDENCE—AND ALSO MIGHT BE EVIDENCE OF A LACK OF REAL PASSION.

I'VE DISCOVERED THAT THOSE INCLINED TO INITIATE BACKUP PLANS TOO QUICKLY OFTEN ARE TRYING TO RATIONALIZE THEIR WAY OUT OF TAKING A RISK IN LIFE OR OUT OF TRULY COMMITTING TO AN AMBITION.

IF YOU FALL FLAT ON YOUR FACE OR GO BANKRUPT, SO WHAT?

EXCEPTIONAL THINKERS KNOW THAT THEY'RE GOING TO FAIL, AND FAILURE DOES NOT CHANGE HOW SMART OR TALENTED THEY ARE, OR HOW MUCH THEIR LOVED ONES LOVE THEM, OR THEIR LONG-TERM POTENTIAL; BANKRUPTCY IS NOT THE END OF THEIR WORLD, BUT JUST ONE MORE CHALLENGE TO SHOW HOW GOOD THEY REALLY ARE. FAILURES DON'T DEFINE US; THE WAY WE RESPOND TO THEM DOES. THIS IS A MOTTO VERY COMMON AMONG THE BEST PER-FORMERS IN THE WORLD.

CONFIDENCE IS NOT YOUR **TRACK RECORD**.

FIRST COMES CONFIDENCE, THEN SUCCESS. PEOPLE WHO BASE THEIR CONFIDENCE ON PAST OR EVEN CURRENT SUCCESSES OFTEN LOSE THEIR SENSE OF DEDICATION AND COMMITMENT. A SENSE OF ENTITLEMENT THEN SETS IN, SO WHY KEEP WORKING HARD? WORSE STILL, BY BASING CONFIDENCE ON YOUR TRACK RECORD, YOU OPEN YOURSELF UP TO BEING EMOTIONALLY CONTROLLED BY THE ROLLER COASTER OF LIFE'S NATURAL UPS AND DOWNS.

BREAK GLASS IN CASE OF LACK OF CONFIDENCE

CONFIDENCE IS NOT A BUTTON TO BE PUSHED.

THERE IS NO GUARANTEED ONE-TO-ONE RELATIONSHIP BETWEEN CONFIDENCE AND SUCCESS. IF EVERY TIME YOU THOUGHT CONFIDENTLY, YOU SCORED, THEN EVERYONE WOULD BE SUPREMELY CONFIDENT.

CONFIDENCE DOES NOT CHANGE YOUR PHYSICAL SKILLS.

IF YOU DO NOT HAVE THE RIGHT SKILLS OR PROPER TRAINING, YOU ARE NOT LIKELY TO SET YOUR FIELD ON FIRE, NO MATTER HOW CONFIDENTLY YOU TRY TO THINK.

CONFIDENCE IS NOT ABOUT **BUILDING SELF-ESTEEM**.

BELIEVING IN YOURSELF IS IMPORTANT, BUT IT'D BETTER BE BASED ON SPECIFIC ACTIONS AND PROCESSES, NOW OR IN THE FUTURE.

I AM AWESOME

CONFIDENCE IS NOT **ARROGANCE**.

REAL ARROGANCE, CALLED SOCIAL ARROGANCE, IS THINKING THAT YOU ARE BETTER THAN OTHER PEOPLE IN GENERAL. CONFIDENCE HAS NOTHING TO DO WITH YOUR WORTH AS A HUMAN BEING, OR WITH A COMPARISON OF YOURSELF TO OTHERS.

ALL BIG CAREERS TAP INTO THE SAME MINDSET: SUPREME CONFIDENCE, ALL YOUR EGGS IN ONE BASKET COMMITMENT, UNREALISTIC DREAMS TO GO WITH YOUR OWN VIEW OF REALITY, AND THE MORE PRESSURE THE BETTER FOR PUTTING YOUR SKILLS AND TALENT ON DISPLAY.

AND WHEN THE GOING GETS TOUGH, THE BEST PERFORMERS WORK LESS; THEIR MINDS ARE FULL OF NOTHING, TOTALLY TRUSTING. THAT'S MY MODEL FOR JOINING THE RANKS OF ALL THOSE OVERACHIEVERS YOU ADMIRE OR ENVY.

I DO NOT HAVE A SECRET FORMULA OR 12-STEP PROGRAM THAT WILL MAKE YOU A SUCCESS. YOU HAVE TO DO THAT.

IF YOU WANT TO RESHAPE YOUR OWN MIND AND FOLLOW SOME OF THE SAME PATTERNS OF THE PLANET'S TOP THINKERS, THE ONLY REAL SECRET IS: IT'S NOT WHAT YOU DO... IT'S HOW YOU DO IT.

OVERACHIEVEMENT IS RIGHT AROUND THE CORNER ...

PLAYING IN THE PRESENT

INTERESTINGLY, SICKNESS HAS BEEN KNOWN TO FOCUS THE MINDS OF THE BEST PERFORMERS IN THE WORLD OF BUSINESS.

THEY FIND A WAY TO TRUDGE OFF TO THE MEETING, WOOZY AND NAUSEATED.

THEY WAKE UP THE MORNING OF THE BIGGEST NEGOTIATION OF THE YEAR WITH A STOMACH BUG. RESCHEDULING MIGHT BLOW THE DEAL; THEY CANNOT AFFORD TO CANCEL. BUT THEY FEEL AWFUL...

TOO SICK TO WANT TO DO ANYTHING BUT FINISH THE MEETING AND GET HOME TO THE SECURITY OF THEIR OWN PRIVATE BATHROOM, THEY SKIP THE USUAL GAMESMANSHIP AND THE BACK-AND-FORTH ONE-UPMANSHIP OVER DETAILS, FOCUSING INSTEAD ON THE MOST SIG-NIFICANT PIECES OF THE NEGOTIATION, ONE AT A TIME, AND—BANG!—THEY CLOSE THE DEAL.

IT IS A STRATEGY THAT WOULD NEVER MAKE IT INTO A HARVARD BUSINESS SCHOOL CASE STUDY, BUT A DEAL GETS SIGNED WITH RECORD-BREAKING EFFICIENCY, ALL BECAUSE ILLNESS INCREASED THEIR **CONCENTRATION**.

PERHAPS YOU HAVE EXPERIENCED THIS YOURSELF.

WHEN CONDITIONS ARE THAT BAD, THE ONLY WAY TO MAKE IT THROUGH IS TO FOCUS INTENTLY ON WHAT YOU HAVE TO DO IN THE IMMEDIACY OF THIS VERY SECOND.

YOUR TOTAL ATTENTION AND ENERGY ARE POURED INTO EXECUTING A TASK DESPITE ALL THE PAIN AND MISERY, LEAVING NO ROOM IN YOUR BRAIN TO THINK ABOUT ANYTHING ELSE, NO ROOM FOR DISTRACTIONS, NO ROOM FOR EXCESS ANALYSIS, NO ROOM FOR ANYTHING BUT SIMPLIFYING THE SITUATION.

FORTUNATELY, CONCENTRATION IS A **VOLUNTARY** ACT.

YOU DON'T HAVE TO GET SICK OR HURT TO ATTAIN INTENSE FOCUS. WHAT YOU NEED IS TO FIND A CENTERPIECE TO YOUR PERFORMANCE—THE ONE OR FEW CORE COMPONENTS THAT ARE THE MOST BASIC AND MOST IMPORTANT.

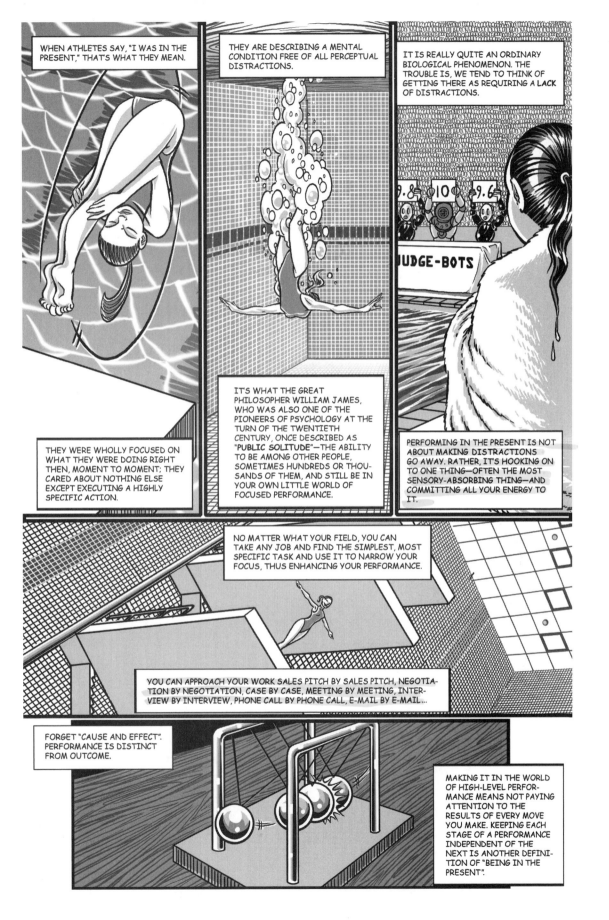

WHEN ATHLETES SAY, "I WAS IN THE PRESENT," THAT'S WHAT THEY MEAN.

THEY ARE DESCRIBING A MENTAL CONDITION FREE OF ALL PERCEPTUAL DISTRACTIONS.

IT IS REALLY QUITE AN ORDINARY BIOLOGICAL PHENOMENON. THE TROUBLE IS, WE TEND TO THINK OF GETTING THERE AS REQUIRING A **LACK** OF DISTRACTIONS.

THEY WERE WHOLLY FOCUSED ON WHAT THEY WERE DOING RIGHT THEN, MOMENT TO MOMENT; THEY CARED ABOUT NOTHING ELSE EXCEPT EXECUTING A HIGHLY SPECIFIC ACTION.

IT'S WHAT THE GREAT PHILOSOPHER WILLIAM JAMES, WHO WAS ALSO ONE OF THE PIONEERS OF PSYCHOLOGY AT THE TURN OF THE TWENTIETH CENTURY, ONCE DESCRIBED AS "**PUBLIC SOLITUDE**"—THE ABILITY TO BE AMONG OTHER PEOPLE, SOMETIMES HUNDREDS OR THOUSANDS OF THEM, AND STILL BE IN YOUR OWN LITTLE WORLD OF FOCUSED PERFORMANCE.

PERFORMING IN THE PRESENT IS NOT ABOUT MAKING DISTRACTIONS GO AWAY. RATHER, IT'S HOOKING ON TO ONE THING—OFTEN THE MOST SENSORY-ABSORBING THING—AND COMMITTING ALL YOUR ENERGY TO IT.

NO MATTER WHAT YOUR FIELD, YOU CAN TAKE ANY JOB AND FIND THE SIMPLEST, MOST SPECIFIC TASK AND USE IT TO NARROW YOUR FOCUS, THUS ENHANCING YOUR PERFORMANCE.

YOU CAN APPROACH YOUR WORK SALES PITCH BY SALES PITCH, NEGOTIATION BY NEGOTIATION, CASE BY CASE, MEETING BY MEETING, INTERVIEW BY INTERVIEW, PHONE CALL BY PHONE CALL, E-MAIL BY E-MAIL...

FORGET "CAUSE AND EFFECT". PERFORMANCE IS DISTINCT FROM OUTCOME.

MAKING IT IN THE WORLD OF HIGH-LEVEL PERFORMANCE MEANS NOT PAYING ATTENTION TO THE RESULTS OF EVERY MOVE YOU MAKE. KEEPING EACH STAGE OF A PERFORMANCE INDEPENDENT OF THE NEXT IS ANOTHER DEFINITION OF "BEING IN THE PRESENT".

JUDGE-BOTS

9.8   10   9.6

COACHES AND SPORT PSYCHOLO-GISTS TELL PITCHERS TO THROW "ONE PITCH AT A TIME"; WE INSTRUCT GOLFERS TO PLAY "ONE SHOT AT A TIME."

BUT DOESN'T ONE PITCH OR GOLF SHOT INFLUENCE THE NEXT ONE?

SURELY, FOR EXAMPLE, IF YOU HIT THE BALL IN THE WOODS INSTEAD OF THE MIDDLE OF THE FAIRWAY, THAT RESULT WILL AFFECT NOT ONLY YOUR NEXT SHOT BUT ALSO YOUR SCORE ON THAT HOLE AND THUS YOUR TOTAL FOR THE ROUND.

HITTING ONE BAD SHOT MAY AFFECT THE NEXT ONE, PHYSICALLY. ERRANT SHOTS CERTAINLY COULD AFFECT THE OVERALL SCORE.

BUT THEY DON'T HAVE TO.

A BAD SHOT CAN JUST AS EASILY LEAD TO A GREAT SHOT—IF YOU ARE THINKING IN THE PRESENT. GREAT PLAYERS DON'T THINK IN TERMS OF HOW A PREVIOUS OUTCOME INFLUENCES THE NEXT; THEY THINK ONLY IN TERMS OF WHAT THEY NEED TO DO TO GIVE THE NEXT OUTCOME ITS BEST POSSIBLE CHANCE.

A PLAYER WHO THINKS EXCEPTIONALLY HITS "ONE SHOT AT A TIME"—AS IF THAT ONE SHOT WAS THE ONLY SHOT HE WOULD HIT ALL WEEK.

THAT'S THE SECOND KEY TO PERFORMING IN THE PRESENT: SEPARATING THE RESULTS FROM THE EXECUTION AND DOING SO INDEPENDENTLY WITH EACH ELEMENT OF YOUR WORK.

TO BE SURE, FEW ACTIVITIES IN LIFE ARE MORE RESULT-ORIENTED THAN BUSINESS TRANSACTIONS.

YET MANAGERS AND THEIR STAFF STILL HAVE TO EXECUTE THE TASKS OF THE DAY. SUCCESSFUL BUSINESSES REQUIRE PEOPLE WHO CAN GET THE JOB DONE WITH SKILL AND PRECISION, EFFECTIVE-NESS, AND EFFICIENCY.

AND AS WE HAVE SEEN, NOTHING DISCOURAGES THE CONCENTRATION NECESSARY TO PERFORM WELL AT GO-TIME MORE THAN WORRYING ABOUT THE OUTCOME, LETTING TASKS PILE TOGETHER OR INFLUENCE ONE ANOTHER, OR FOCUSING ON THE PAST OR THE FUTURE.

YOU ARE HERE

MASTERING THE ART OF BEING IN THE PRESENT CAN BE AN EXTRAORDINARY TOOL FOR BUSINESS SUCCESS.

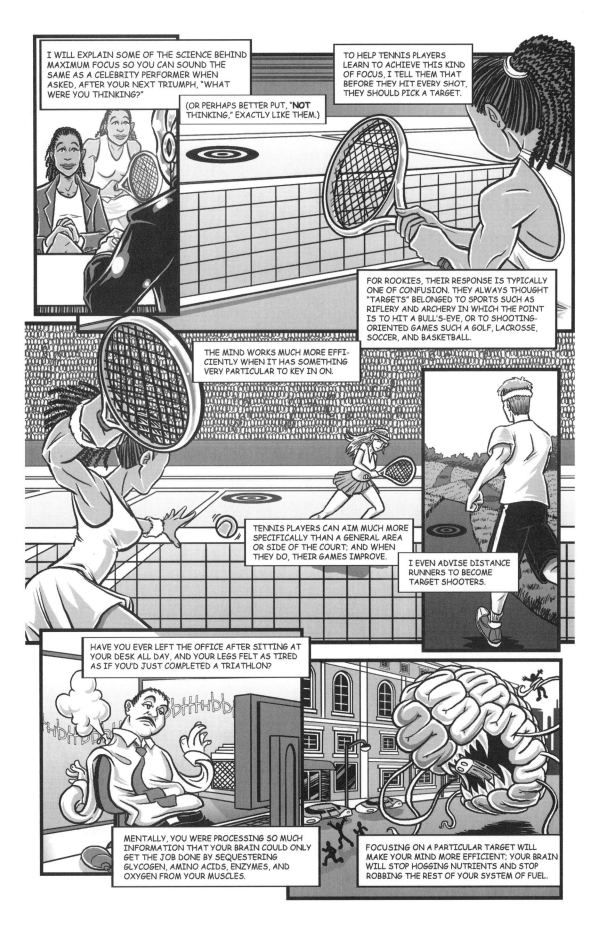

IF YOU APPROACH YOUR WORK WITH INCREASED FOCUS, YOU'LL PREVENT MUSCLE CATABOLISM.

OF COURSE, THE POINT ISN'T ACTUALLY THE PHYSICAL HITTING OF A SPOT. IT'S ABOUT HAVING SOMETHING TO FILL YOUR MIND WITH, VIVIDLY AND COMPLETELY.

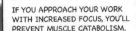

focus

SUCCESS IN TARGET SHOOTING IS ALL ABOUT WHAT HAPPENS **BEFORE** YOU PULL THE TRIGGER.

WHAT DO I MEAN WHEN I SAY "PICK A TARGET"?

JUST THAT, PICK **ONE**—GET YOUR EYES (OR BRAIN) TO SELECT THE FINEST, MOST DETAILED, MOST IMMEDIATE ELEMENT OF YOUR PERFORMANCE, AND THEN SIMPLY REACT TO WHAT YOUR EYES SEE.

FOR MOST BUSINESSPEOPLE, "TARGET" MEANS AN OUTCOME, SOMETHING TO SHOOT FOR IN SALES, EARNINGS, OR PROFITS.

Previous Purchases

Current Needs

Contract

Presentation

Close

Referrals

Lunch Meet

Possible Objections

BUT OUTCOMES ARE A DISTRACTION FROM TOTAL AND REAL CONCENTRATION. IN BUSINESS-RELATED PERFORMANCE, YOUR TARGET SHOULD BE A KEY IN THE **PROCESS** OF WHAT YOU ARE DOING.

IN SALES, FOR EXAMPLE, THE TARGET IS NOT MAKING THE SALE BUT DOING WHAT IS MOST LIKELY TO LEAD TO THE SALE.

2031 Qtr. 4 SALES

ONCE YOU'VE SELECTED A FOCUSED, PREPARATION AND RESEARCH-BACKED TARGET, EXECUTE IT WITHOUT RESERVATION. IF YOU START SECOND-GUESSING YOUR STRATEGY AND SWITCH TO ANOTHER IN THE MIDDLE OF THE SALE, YOU WILL HAVE TAKEN YOUR MIND'S EYE OFF THE BALL.

3

CREDIT 00

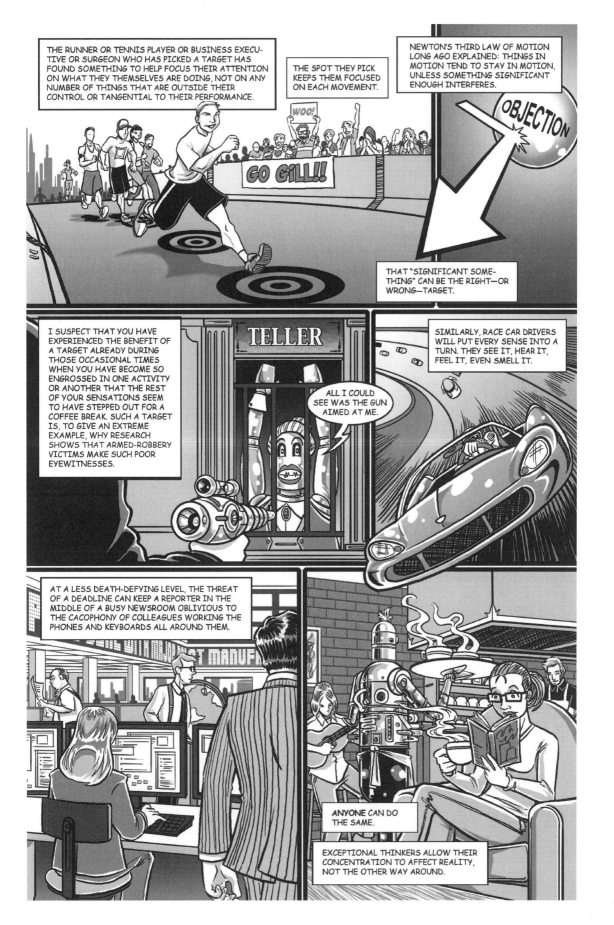

43

THERE IS A DIFFERENCE BETWEEN PUTTING YOUR EFFORT INTO THINKING ABOUT EVERYTHING AFFECTING YOUR PERFORMANCE —ALL THE DISTRACTIONS, EVALUATIONS, AND CONSEQUENCES— AND PUTTING THE SAME EFFORT INTO THINKING ABOUT THE PERFORMANCE ITSELF.

SUPPOSE AN ELEPHANT DROPPED DEAD IN THE MIDDLE OF YOUR LIVING ROOM. THE MORE YOU TRY TO SNEAK PAST THAT ELEPHANT, TO STEP AROUND IT, OR TO IGNORE IT ALTOGETHER, THE MORE IT SITS THERE ROTTING AND STINKING; IT GETS MORE AND MORE NOTICEABLE, AND EVEN HARDER TO AVOID.

THE SAME THING HAPPENS WHEN YOU TRY NOT TO THINK ABOUT SOMETHING. IT WILL GET INTO YOUR HEAD MORE FIERCELY!

THE SOLUTION IS NOT TO **REMOVE** A THOUGHT, IMAGE, OR FEELING FROM YOUR BRAIN, BUT TO SUMMON UP A NEW ONE TO **REPLACE** IT.

THE MORE ABSORBED YOU ARE IN AN ALTERNATIVE TARGET, THE MORE THE UNPLEASANT OR PERFORMANCE-HAMPER-ING INFORMATION GETS PUSHED OUT.

OF COURSE, TRYING TO ACHIEVE INTENSE, TARGET-SPECIFIC CONCENTRATION MAY NOT BE EASY INITIALLY. DO NOT EXPECT TO BE AN EXCEPTIONAL THINKER THE FIRST TIME YOU TRY TO GET THERE.

YOU CAN MASTER FOCUS MORE READILY WHEN YOU THINK OF IT AS A CHOICE YOU CAN ACTUALLY MAKE.

DECIDE WHAT YOUR TARGET IS **BEFORE** YOU PERFORM; MAKE IT SINGULAR, SPECIFIC, PROCESS-ORIENTED, VIVID. AND GET INTO THAT KIND OF INTENTIONAL CONCENTRATION ROUTINELY.

PRE-PERFORMANCE ROUTINES

ANYONE INTENT ON BECOMING A TOP PERFORMER WILL NEED A **PRE-PERFORMANCE ROUTINE**.

AS PERFORMERS, WE HAVE TO DIVIDE OUR LIVES IN HALF BETWEEN PREPARING AND PERFORMING—BETWEEN FACING PRESSURE AND GETTING READY TO FACE PRESSURE.

A ROUTINE HELPS YOU MAKE THE CRITICAL TRANSITION.

THE DOCTOR MOVES FROM PRE-OP TO SURGERY.

THE MANAGER MOVES FROM CREATING A PRESENTATION TO DELIVERING IT.

IN EVERY CAREER, WE HAVE TO SHIFT FROM ORDINARY WORKDAY ACTIVITIES TO MOMENTS WHEN WE ARE REQUIRED TO EXECUTE TO THE BEST OF OUR ABILITY.

OF COURSE, WE'RE ALL HUMAN. WE EXPERIENCE OCCASIONS WHEN SOMETHING BREAKS OUR CONCENTRATION, FOR ONE REASON OR ANOTHER, AND WE HAVE TO FIND A WAY BACK INTO THE PERFORMANCE MODE.

CONSIDER THAT THE SURGEON MIGHT HAVE HAD AN ARGUMENT WITH HIS SPOUSE MINUTES BEFORE A SCHEDULED OPERATION.

HOW DOES HE ALTER THOSE INTENSE FEELINGS IN ORDER TO DEAL WITH THE PATIENT ON THE OPERATING TABLE?

MOST PEOPLE DO NOT REALIZE THAT THEY ALREADY POSSESS SOME ELEMENTS OF A PRE-PER-FORMANCE ROUTINE.

PRE-PERFORMANCE ROUTINES SHOULD BE DESIGNED TO GET YOU TO **THINK** CLEARLY AND SIMPLY DURING AN UPCOMING EVENT—TO BE CONFI-DENT, TO FOCUS, TO TAKE ADVANTAGE OF THE PHYSICAL RESPONSE A PRESSURE SITUATION SPARKS.

MOST PEOPLE DON'T REALIZE THAT THEY PURPOSEFULLY CAN CHANGE THEIR MOOD OR HOW THEY THINK. SO THEY LEAVE IT TO CHANCE. YOU DON'T NEED TO DO THAT.

EXCEPTIONAL PERFORMERS USE PRE-PERFORMANCE ROUTINES TO GET THEIR MINDS INTO THE RIGHT CONDITION TO ALLOW ALL THEIR TALENT AND YEARS OF PRACTICE TO DO THEIR THING.

A ROUTINE IS NOTHING MORE THAN A SYMBOLIC GESTURE, A MENTAL AND/OR PHYSICAL EXERCISE TO GET INTO THE PRESENT, FREE OF EVALUATIONS.

THAT LITTLE ROUTINE IS A TOOL FOR MAKING THE TRANSITION BETWEEN TRAINING AND TRUSTING—TWO OPPOSING MINDSETS.

THE WORST THING YOU CAN DO IS LET YOUR ROUTINE DOMINATE YOUR PERFORMANCE.

THE PURPOSE OF A ROUTINE IS TO MOVE SMOOTHLY INTO PERFORMANCE, NOT TO CHECK ONE THING OFF AND THEN CHECK THE NEXT.

STEP 1
STEP 2
STEP 3
STEP 4
STEP 9
STEP 5
STEP 6
STEP 8
STEP 7
STEP 16
STEP 10
STEP 12
STEP 11
STEP 13
STEP 14
STEP 15

THINK OF IT AS A LOT LESS LIKE A COUNTDOWN AND A LOT MORE LIKE A WARM-UP DANCE.

TO GET THE RIGHT FEELING GOING YOU MIGHT NEED TO CHANGE YOUR ROUTINE OCCASIONALLY.

ASIDE FROM RELAXATION, THE MOST OVERUSED (AND OFTEN **OVERRATED**) ROUTINE IN PSY-CHOLOGY IS **VISUALIZATION**.

WARNING! WARNING! DANGER!

CONTRARY TO POP PSYCHOLOGY, VISUALIZING YOUR PERFORMANCE AHEAD OF TIME CAN ACTUALLY BE A DETERRENT TO MANY PERFORMERS WHO, BEING HUMAN, ARE FAR FROM PERFECT. WHEN THE PERFORMANCE THAT COMES OUT DOES NOT MATCH THE PERFORMANCE IN YOUR MIND, YOU'RE THROWN RIGHT BACK INTO THE TRAINING MINDSET, COMPAR-ING, OVER-EVALUATING, ANALYZING WHERE YOU WENT WRONG.

PHILOSOPHIES

SOME EXAMPLES MIGHT INCLUDE:

I DON'T KNOW WHAT IT WILL BE OR WHEN, BUT I WILL LEARN A TON AND END UP IN A GREAT NEW PLACE.

EVERYBODY PUTS ON HIS OR HER PANTS ONE LEG AT A TIME.

NO MATTER WHAT I AM DOING, I'LL FIND A FUN WAY TO DO IT.

WORST PERFORMANCE EVER

LIFE, OF COURSE, CANNOT ALWAYS BE FUN. THE BEST PERFORMERS, HOWEVER, ALWAYS HAVE STRATEGIES ON HAND TO HELP THEM COPE WITH TOUGH TIMES, EVEN WITH TRAGEDY, IN A MANNER THAT PROMOTES GROWTH AND LOVE.

I CANNOT CONTROL EVENTS, BUT I CAN CONTROL MY REACTION TO THEM.

I'M SORRY BUT I CANNOT HAND OUT READY-TO-WEAR PHILOSOPHIES.

EZ PHILOSOPHY CDs $3.00

the DOCTOR is IN

THAT'S PART OF BEING AN EXCEPTIONAL THINKER—FIGURING OUT YOUR CORNERSTONE PRINCIPLES. THEY ARE YOUR GUIDING THOUGHTS, YOUR FOUNDATION MOTTO FOR CONSISTENCY IN YOUR APPROACH AND EXECUTION, SO THEY NEED TO BE PERSONAL TO YOU, IN YOUR OWN WORDS.

EVERY PERFORMANCE IS A NEW CHANCE TO SHOW WHAT I CAN DO, TO TEST MY THINKING, TO TEST HOW MUCH FUN I CAN HAVE BEING IN THE MOMENT.

AN EFFECTIVE PHILOSOPHY OF PERFORMANCE SHOULD BE SIMPLE AND UNAMBIGUOUS.

PERFORMERS USUALLY HAVE JUST ONE, MAYBE TWO GUIDING PRINCIPLES AS THEY ENTER EVERY PERFORMANCE SITUATION. OTHERWISE THEY HAVE TOO MUCH TO THINK ABOUT AND WILL BE FORCED TO CONCENTRATE MORE ON THEIR PHILOSOPHY THAN ON PERFORMING.

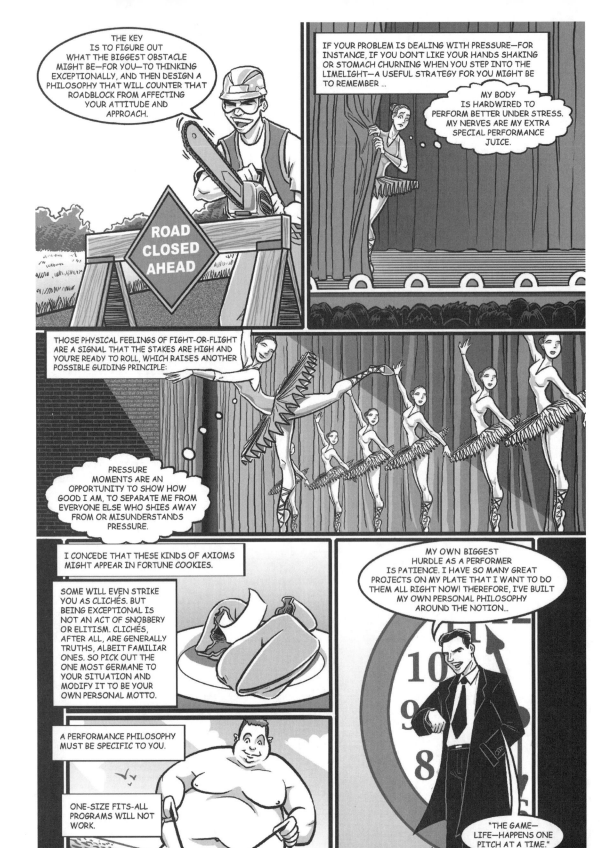

MOST PEOPLE DO NOT HAVE A DELIBERATELY CRAFTED, OVERRIDING SET OF THOUGHTS TO USE AS THEY GO ABOUT EXECUTING THEIR DAILY STRATEGIES AND PLANS.

THEY DEFINITELY HAVE THEIR GOALS, "SUCCESS FORMULAS", AND BUSINESS PLANS. BUT THOSE ARE THE X'S AND O'S OF A CAREER. AND AS ANY FAMOUS COACH WILL TELL YOU, CHAMPIONSHIPS ARE NOT WON ON X'S AND O'S ALONE.

A GREAT PHILOSOPHY OF PERFORMANCE IS A PRACTICAL WAY OF KEEPING THE "INTANGIBLE" OBSTACLES—DISTRACTIONS, FIRE DRILLS, UNPREDICTABILITY, NEGATIVE THOUGHTS THAT JUST POP INTO YOUR HEAD, OCCURRENCES OUT OF YOUR CONTROL—FROM GETTING IN THE WAY OF WELL PLANNED, WELL PREPARED X'S AND O'S.

FOR STARTERS, HERE ARE SOME SAMPLE SUGGESTIONS FOR PERFORMANCE PHILOSOPHIES TO THRIVE UNDER PRESSURE.

PRESSURE IS WHAT TURNS COAL INTO DIAMONDS.

PRESSURE MOMENTS ALLOW ME TO USE ALL OF MY TRAINING TO DO SOMETHING TRULY WORTHWHILE.

THE ABSENCE OF FEELING PRESSURE WOULD BE A SURE SIGN OF BOREDOM, DISINTEREST, AND A LACK OF IMPORTANCE IN WHAT YOU DO.

YOUR OBSTACLE MIGHT NOT BE PRESSURE AT ALL. IT COULD BE, FOR INSTANCE, SPENDING TOO MUCH TIME PREPARING FOR THINGS RATHER THAN ACTUALLY DOING THEM. IN THIS CASE YOU MIGHT TRY YOUR OWN VARIATIONS OF THESE AXIOMS...

"YOU WILL NEVER WIN A COMPETITION IN PRACTICE."

"I CANNOT BE SUCCESSFUL IN A JOB BY STAYING IN SCHOOL OR CONTINUALLY ATTENDING TRAINING SEMINARS."

"WORKING EVEN HARDER IS LIKE ENTERING A LONG-DIVISION COMPETITION WITH ONLY PENCIL AND PAPER—YOU'LL GET WHOOPED BY SOMEONE WITH A CALCULATOR."

NEGATIVE THINKING IS EXTREMELY POWERFUL.

IF YOU BELIEVE CATASTROPHE OR EVEN A SMALL HICCUP IS ONE STEP AROUND THE CORNER, THEN IT PROBABLY WILL BE. OUR MIND IS DARN GOOD AT TURNING NEGATIVE THOUGHTS INTO REALITY.

THE OTHER WAY AROUND, "THINK POSITIVELY" IS GENERALLY GOOD ADVICE. BUT GUESS WHAT? IT ALSO CAN SOMETIMES BE AN OBSTACLE TO EXCEPTIONAL THINKING.

EXCEPTIONAL THINKERS TEACH THEMSELVES THE POWER OF POSITIVE *ACTION*. THEY DON'T STOP TO GIVE THEMSELVES A PEP TALK OR THINK ABOUT HOW GREAT THE ACT IS GOING TO BE. INSTEAD, THEY ACT.

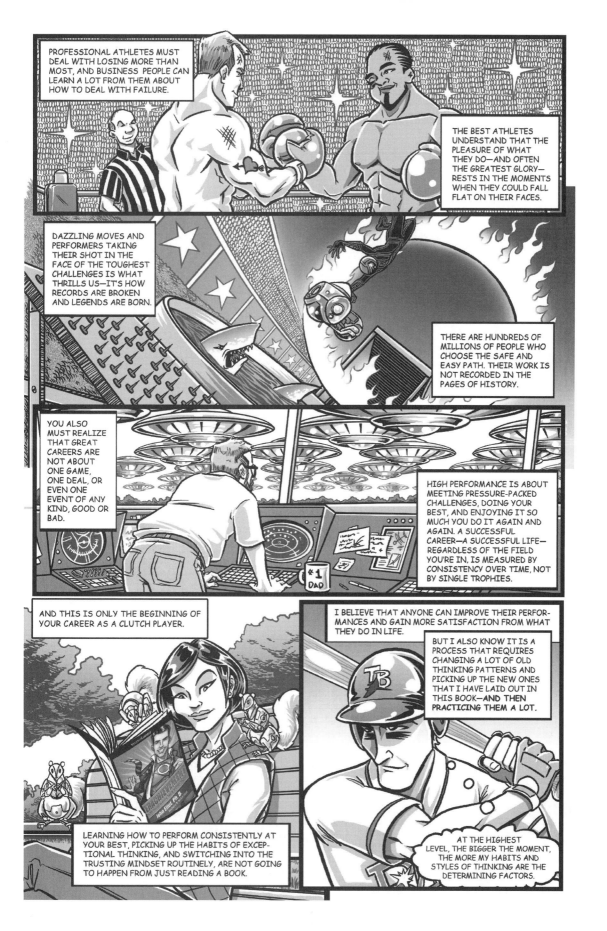

PROFESSIONAL ATHLETES MUST DEAL WITH LOSING MORE THAN MOST, AND BUSINESS PEOPLE CAN LEARN A LOT FROM THEM ABOUT HOW TO DEAL WITH FAILURE.

THE BEST ATHLETES UNDERSTAND THAT THE PLEASURE OF WHAT THEY DO—AND OFTEN THE GREATEST GLORY—RESTS IN THE MOMENTS WHEN THEY COULD FALL FLAT ON THEIR FACES.

DAZZLING MOVES AND PERFORMERS TAKING THEIR SHOT IN THE FACE OF THE TOUGHEST CHALLENGES IS WHAT THRILLS US—IT'S HOW RECORDS ARE BROKEN AND LEGENDS ARE BORN.

THERE ARE HUNDREDS OF MILLIONS OF PEOPLE WHO CHOOSE THE SAFE AND EASY PATH. THEIR WORK IS NOT RECORDED IN THE PAGES OF HISTORY.

YOU ALSO MUST REALIZE THAT GREAT CAREERS ARE NOT ABOUT ONE GAME, ONE DEAL, OR EVEN ONE EVENT OF ANY KIND, GOOD OR BAD.

HIGH PERFORMANCE IS ABOUT MEETING PRESSURE-PACKED CHALLENGES, DOING YOUR BEST, AND ENJOYING IT SO MUCH YOU DO IT AGAIN AND AGAIN. A SUCCESSFUL CAREER—A SUCCESSFUL LIFE—REGARDLESS OF THE FIELD YOU'RE IN, IS MEASURED BY CONSISTENCY OVER TIME, NOT BY SINGLE TROPHIES.

AND THIS IS ONLY THE BEGINNING OF YOUR CAREER AS A CLUTCH PLAYER.

I BELIEVE THAT ANYONE CAN IMPROVE THEIR PERFOR-MANCES AND GAIN MORE SATISFACTION FROM WHAT THEY DO IN LIFE.

BUT I ALSO KNOW IT IS A PROCESS THAT REQUIRES CHANGING A LOT OF OLD THINKING PATTERNS AND PICKING UP THE NEW ONES THAT I HAVE LAID OUT IN THIS BOOK—AND THEN PRACTICING THEM A LOT.

LEARNING HOW TO PERFORM CONSISTENTLY AT YOUR BEST, PICKING UP THE HABITS OF EXCEP-TIONAL THINKING, AND SWITCHING INTO THE TRUSTING MINDSET ROUTINELY, ARE NOT GOING TO HAPPEN FROM JUST READING A BOOK.

AT THE HIGHEST LEVEL, THE BIGGER THE MOMENT, THE MORE MY HABITS AND STYLES OF THINKING ARE THE DETERMINING FACTORS.

ABOUT THE AUTHOR

Dr. John Eliot, descendant of Harvard president, Charles Eliot, and Nobel Prize winner, T.S. Eliot, is a decorated professor of management and performance psychology. He currently consults for Stanford and Texas A&M University, and has held faculty leadership positions at Rice and the SMU School of Business. When not inspiring our next generation, Dr. Eliot is an advisor to the stars. Clients have included Microsoft, Yahoo, Sony, Shell, Goldman Sachs, NASA, the Texas Medical Center, the United States Olympic Committee, the Tampa Bay Rays, and hundreds of CEOs, surgeons, professional athletes, and performing artists. Dr. Eliot also serves as the chairman of GET IN THE BOX! (www.getinthebox.org), a national non-profit foundation tackling the problems of youth obesity and diabetes.

ABOUT THE ARTIST

Nathan Lueth has been a lover of comics and animation since birth and an artist for even longer. He decided to make professional illustration his career after graduating the Minneapolis College of Art and Design and deciding he didn't want to deliver pizza any more. Conveniently both happened on the same day.

Nathan currently resides in St. Paul, Minnesota, with his imminent wife, Nadja, two cats, turtle, and action figure collection. His list of clients includes the likes of Target, General Mills, Stone Arch Books, and Frivolous Entertainment. His work can be also be seen in the hit webcomic *Impure Blood*, and the October 2011 Round Table Comics release of Alesia Shute's *Everything's Okay*.

Strategic Vision

The vision is to strategically align the Foundation with organizations, facilities, and other interested parties to help create awareness AND a sense of urgency to improve the lifestyle of our youth. Join hands with us – a coalition of professional athletes, executives and professors who are saying "ENOUGH!" It's time to dig in, get in the batter's box, and take responsibility for the health of our children!

Privately funded solutions in lieu of public bureaucracy:

GET IN THE BOX channels capital from donors to non-profit corporations and leaders. The GET IN THE BOX Foundation answers to their donors, ensuring that capital allocations and investment results meet donor expectations.

Two issues: calories in (food) and calories out (activity)

• Food: The GET IN THE BOX Foundation works both directly and indirectly through 3rd party non-profits to help families make educated food choices and lead healthier lives.

• **Activity:** The GET IN THE BOX Foundation will attack inertia and inactivity by building a steady capital base, powered by donors inspired to make a difference. Thoughtfully partnering with non-profits with leading strategies and proven results in driving youth activity.

GET IN THE BOX Foundation's mission is to provide funding for improvement to the physical and mental well-being of American youth.

Learn more at www.getinthebox.org
Find us on Twitter and Facebook

Look for these other titles from SmarterComics and Writers of the Round Table Press:

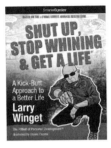

Shut Up, Stop Whining & Get a Life from SmarterComics
by Larry Winget Illustrated by Shane Clester

Internationally renowned success philosopher, business speaker, and humorist, Larry Winget offers advice that flies in the face of conventional self-help. He believes that the motivational speakers and self-help gurus seem to have forgotten that the operative word in self-help is "self." That is what makes this comic so different. *Shut Up, Stop Whining & Get a Life from SmarterComics* forces all responsibility for every aspect of your life right where it belongs—on you. For that reason, this book will make you uncomfortable. Winget won't let you escape to the excuses that we all find so comforting. The only place you are allowed to go to place the blame for everything that has ever happened to you is to the mirror. The last place most of us want to go.

Think and Grow Rich from SmarterComics
by Napoleon Hill Illustrated by Bob Byrne

Think and Grow Rich has sold over 30 million copies and is regarded as the greatest wealth-building guide of all time. Read this comic version and cut to the heart of the message! Written at the advice of millionaire Andrew Carnegie, the book summarizes ideas from over 500 rich and successful people on how to achieve your dreams and get rich doing it. You'll learn money-making secrets - not only what to do but how - laid out in simple steps.

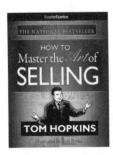

How to Master the Art of Selling from SmarterComics
by Tom Hopkins Illustrated by Bob Byrne

With over one million copies sold in its original version, *How to Master the Art of Selling from SmarterComics* motivates and educates readers to deliver superior sales. After failing during the first six months of his career in sales, Tom Hopkins discovered and applied the very best sales techniques, then earned more than one million dollars in just three years. What turned Tom Hopkins around? The answers are revealed in *How to Master the Art of Selling from SmarterComics*, as Tom explains to readers what the profession of selling is really about and how to succeed beyond their imagination.

The Art of War from SmarterComics
by Sun Tzu Illustrated by Shane Clester

As true today as when it was written, *The Art of War* is a 2,500-year-old classic that is required reading in modern business schools. Penned by the ancient Chinese philosopher and military general Sun Tzu, it reveals how to succeed in any conflict. Read this comic version, and cut to the heart of the message!

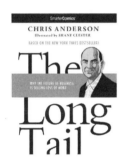

The Long Tail from SmarterComics
by Chris Anderson Illustrated by Shane Clester

The New York Times bestseller that introduced the business world to a future that's already here. Winner of the Gerald Loeb Award for Best Business Book of the Year. In the most important business book since *The Tipping Point*, Chris Anderson shows how the future of commerce and culture isn't in hits, the high-volume head of a traditional demand curve, but in what used to be regarded as misses—the endlessly long tail of that same curve.

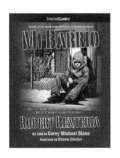

Mi Barrio from SmarterComics
by Robert Renteria as told to Corey Michael Blake
Illustrated by Shane Clester

"Don't let where you came from dictate who you are, but let it be part of who you become." These are the words of successful Latino entrepreneur Robert Renteria who began life as an infant sleeping in a dresser drawer. This poignant and often hard-hitting comic memoir traces Robert's life from a childhood of poverty and abuse in one of the poorest areas of East Los Angeles, to his proud emergence as a business owner and civic leader today.

For more information, please visit www.smartercomics.com

Overachievement and other SmarterComics™ books
are available for download on the iPad and other devices.

www.smartercomics.com

The book that inspired the comic...

Available everywhere books are sold.

SmarterComics™

www.smartercomics.com